Written by Nyree Bevan
Cover design by Phillip Colhouer

© 2020 Jenny Phillips
goodandbeautiful.com

Did you know that in your backyard there is an ecosystem? You might be wondering what that is. An ecosystem is a community of living organisms that interact in a particular place. The organisms all work together to create a circle of life. Our incredible earth is full of many ecosystems, and they are all connected. In fact, the earth is one gigantic ecosystem. Yet some are small, like your neighborhood or your own backyard. Even just a small potted plant is an ecosystem!

Let's travel around the world and explore the ecosystems of three different backyards.

First, we'll head to beautiful Dunsborough in

WESTERN AUSTRALIA

KOOKABURRA

Right next to a river, there is a little green clapboard house, and in the backyard are lots of tall trees. Up in the trees live many wild birds, such as weiros, willie wagtails, and kookaburras, creating quite a ruckus. Down in the grass, you might see kangaroos bounding away or snakes seeking some shade from the hot sun. All year long, pesky mosquitoes buzz around.

DIAMOND PYTHON

The weiros like to eat the seeds and berries in the trees and bushes. In turn, they help to spread those seeds so more bushes and trees can grow.

WEIRO

WILLIE WAGTAIL

Willie wagtails eat mosquitoes, which is surely a great blessing to the other animals and people in the area! The kookaburras eat insects, fish from the river, and even small snakes.

KANGAROO

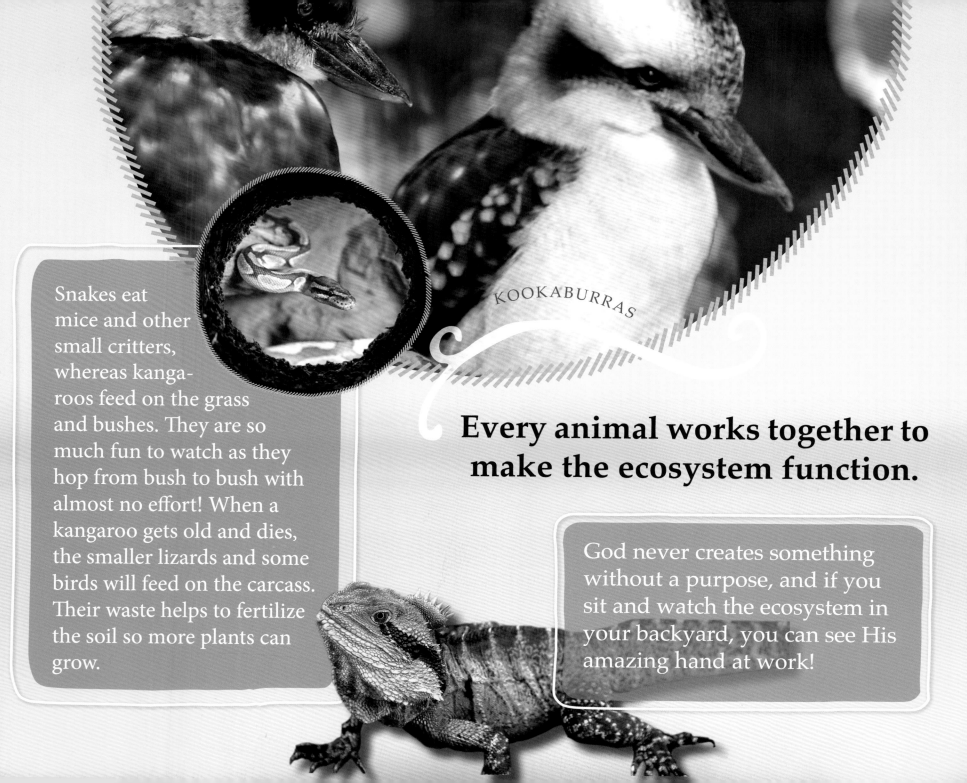

KOOKABURRAS

Snakes eat mice and other small critters, whereas kangaroos feed on the grass and bushes. They are so much fun to watch as they hop from bush to bush with almost no effort! When a kangaroo gets old and dies, the smaller lizards and some birds will feed on the carcass. Their waste helps to fertilize the soil so more plants can grow.

Every animal works together to make the ecosystem function.

God never creates something without a purpose, and if you sit and watch the ecosystem in your backyard, you can see His amazing hand at work!

Kookaburras catch snakes and beat them on rocks or impale them on tree branches, killing the snakes before eating them.

Kookaburras are also known as laughing kookaburras because their call sounds like a laughing monkey.

Making their nests in tree holes, kookaburras usually have two to four young at a time. The young often stay with their parents to help raise the next year's babies.

Kookaburras eat venomous snakes in Australia.

KING BROWN SNAKE

Willie wagtails are named because of the repeated side-to-side twitching, or wagging, of their tails.

The eggs of willie wagtails are cream colored with brown and gray specks. They are incubated by both the mom and dad.

WILLIE WAGTAILS CAN EAT THEIR OWN WEIGHT IN INSECTS EVERY DAY!

Weiros are called cockatiels in the United States.

Weiros usually live an average of 15 years in the wild but can live up to 25 years or more!

A species of parrot, the weiro can be hand raised from an egg and often can learn to talk or sing various melodies.

Kangaroos can jump over a fence 1.8 meters (6 feet) tall in one bound and move at speeds of up to 56 kilometers (35 miles) per hour!

Kangaroos travel together in groups called mobs and are led by the largest male, called a boomer.

Some species of kangaroo can grow up to 2.1 meters (6.9 feet) tall and weigh up to 90.7 kilos (200 pounds)!

A MOB OF KANGAROOS

CARPET PYTHON

THE CARPET PYTHON CAN BE BETWEEN 1.8 AND 4 METERS (6–13 FEET) LONG AND WEIGH MORE THAN 13.6 KILOS (30 POUNDS).

TIGER SNAKE

Tiger snakes are common near Dunsborough, and they can often be seen on bicycle paths. But it's wise to stay out of their way because they are extremely venomous!

The desert death adder, which is found all over Western Australia, uses a trick to attract its prey. The end of its tail looks a bit like a worm, so it hides in bushes and only sticks out its tail, wiggling it like a worm. When prey like lizards, skinks, or water dragons try to catch the "worm," they get eaten by a death adder instead!

DESERT DEATH ADDER

On the other side of our planet in the lush, green state of Maryland, USA, there is a place called Germantown.

MARYLAND, USA

Here we find many houses, cars, people, and neighborhoods. Behind a brown town house, there is a small woodland with a creek flowing through it. The trees and plants give the community fresh air to breathe, and there are paths where people like to walk.

NORTH AMERICA

BALD EAGLE

The bald eagle isn't actually bald. In English, the word "bald" used to mean "white," so it gets its name from the distinctive white feathers on its head.

There are so many trees and flowers in the woods. It is nice and cool under the canopy of leaves, and you can often hear creatures moving about.

A TRILLIUM PLANT CAN LIVE UP TO 25 YEARS!

TRILLIUM

BLACK-EYED SUSAN

WHITE-TAILED DEER FAWN

If you walk down the path from the backyard of the brown town house, you can see an old maple tree near the creek that was once struck by lightning. Half of the tree is black and dead, and half of the tree is green and alive. Making a nest in a burned-out cavity of this maple tree is an eastern screech owl, only 18 centimeters (7 inches) tall. That old maple also provides shade for the smaller plants around it.

The complex patterns of spots and markings give the eastern screech owl great camouflage.

SCREECH OWL

Eastern screech owls are quite small, about the size of a robin, but they have a big, spooky trill that can give you quite a fright on a dark night.

When there is plenty of food, the eastern screech owl will put away three to four days' worth of food next to its nest.

Old tree branches from the maple tree fall to the ground during the winter and summer storms and begin to decay. Lichens and moss grow on the logs, nourished by the decomposing wood, and in turn they help maintain cool, moist conditions that allow the wood to continue to decay.

LICHENS

Lichen is helpful to humans; it is used as a food source and for medicinal purposes.

TRUMPET LICHENS

Lichen is formed through a symbiotic relationship between algae or cyanobacteria and fungi.

BUBBLE BRAID LICHENS

Fungi, such as mushrooms, use the nutrients from these old branches to grow. As these fungi grow, they produce enzymes that break down the log.

Carpenter ants, beetles, and pill bugs feed off the mushrooms and decomposing wood. Small animals such as meadow voles come to eat the insects.

CARPENTER ANT

Carpenter ants are different from termites. Carpenter ants do not eat wood, but they do a lot of damage to wood by hollowing it out for nesting.

Often there are two different colonies of carpenter ants: the parent colony, which contains a queen that lays eggs, a brood, and several worker ants; and the satellite colony, which only has worker ants.

BEETLE

THERE ARE MORE THAN 30,000 DIFFERENT KINDS OF BEETLES IN THE UNITED STATES.

Beetles are beneficial to humans, especially as predators who eat insects that harm crops and trees.

The tail of a meadow vole can be $1/3$ to $1/2$ the length of its body.

Meadow voles are herbivores and eat seeds, leaves, stems, and roots. They store roots and tubers in their burrows to eat during winter.

A MEADOW VOLE'S GESTATION IS 20 TO 21 DAYS, AND IT CAN PRODUCE UP TO 17 LITTERS IN ONE YEAR.

YOUNG BANK VOLE

MEADOW VOLE

Both male and female kestrels have two distinct vertical lines on the sides of their faces. These are sometimes called a "mustache" or "sideburns."

Parent kestrels will often hunt with their young, giving them a chance to learn to hunt before having to do it on their own.

With all the voles, mice, and insects running around, these woods are the perfect hunting ground for American kestrels.

There are crawdads and fish in the creek, which some of the birds eat. The creek provides enough water to keep tree roots moist and growing strong.

Crawdads are also called crayfish, crawfish, and crawdaddies, depending on where you are from.

THE LARGEST CRAWDADS CAN BE OVER 38 CM (15 INCHES) AND WEIGH 3.6 KILOS (8 POUNDS)!

CRAWDADS CAN BE A WIDE VARIETY OF COLORS, INCLUDING GREEN, RED, SANDY YELLOW, AND DARK BROWN.

The creek also provides a place for deer, raccoons, squirrels, foxes, and many other animals to drink.

Each fall the old maple tree's leaves turn from green to yellow and finally orange. The maple leaves fall and scatter on the ground as the nearby oak trees drop acorns onto the colorful leaves below.

With cheeks that can expand up to three times the size of its head, a chipmunk can store a lot of food in its mouth to take back to its den for winter.

Did you know a single chipmunk can gather as many as 165 acorns in one day?

Chipmunks make dens for their litters, but they also have been known to inhabit hollow logs or even old bird nests to care for their young.

Leaves change color because they stop their food-making process due to the length of daylight and changes in temperature.

Under the blanket of leaves, the pill bugs and slow-moving worms come up to feed. They help break down the old leaves to provide nutrients for the trees to keep growing.

PILL BUG

PILL BUGS, ALSO KNOWN AS SLATERS OR ROLY-POLIES, HAVE 14 SEGMENTS, ALLOWING THEM TO ROLL INTO A TIGHT BALL. THE MOTHERS KEEP THEIR EGGS IN A POUCH ON THE UNDERSIDE OF THEIR BODIES.

AN EARTHWORM CAN LIVE UP TO 6 YEARS IN THE WILD!

PILL BUG IN A DEFENSIVE POSITION

Germantown used to be little more than farms and a train station. Now it has grown into a large suburb and is a bustling business area. However, the community has allowed some areas to remain woods to help preserve wild space, such as Seneca Creek State Park, pictured here. This is an important part of keeping ecosystems alive.

There are many lakes in Maryland, but none of them are natural lakes; they have all been made by people to help preserve the water from the rains.

This ecosystem is a great example of the amazing work of our Creator!

All the animals and plants work together to create a healthy and beautiful place.

What about people who live in large urban areas with buildings and cars everywhere?

How do you find an ecosystem when you don't even have a yard?

ASIA

HONG KONG

On the 20th floor of a high-rise apartment building in Hong Kong lives a lovely couple and their young daughter. They have a very small patio that looks out over the city.

On their little patio, they have three boxes. One box has green vegetables such as lettuce and kale. Next to this is a box that has small cucumbers. Most interesting is the final box, which is filled with lots of wilted vegetables and bits of torn paper.

These items break down in the soil and feed the many worms that also live in the box. Sometimes, in the morning after rain, you might see a Eurasian blackbird tugging one of these worms out of the dirt to take home for breakfast. Or on a warm afternoon, you might hear the buzzing of bees coming to pollinate the cucumber blossoms.

A MALE EURASIAN BLACKBIRD IS ALL BLACK EXCEPT FOR ITS YELLOW BEAK AND THE YELLOW RING AROUND ITS EYES.

The happy, flute-like song of the Eurasian blackbird can be heard from winter to summer, most often early in the morning and evening.

THERE ARE TWO COMMON TYPES OF BEES IN HONG KONG: THE ASIATIC HONEY BEE AND THE ITALIAN BEE.

The Asiatic honey bee is one of the most physically varied species of honey bee.

THE ITALIAN BEE IS KNOWN FOR ITS GENTLE NATURE AND ABILITY TO EASILY ADAPT TO DIFFERENT CLIMATES.

The family eats most of the vegetables, and what they don't eat goes to feed the worms. As the worms eat the leftovers, they leave behind waste, also known as castings, which make some of the best fertilizer you can find. The family members collect the castings from the worms to put into the soil of their vegetable boxes. This helps feed the plants and produce large, tasty vegetables.

Vermiculture is the farming of worm castings. You can create your own worm casting farm with your parents' permission. It is a great way to help compost the uneaten food from your table and help the ecosystem in your yard.

Worms are a very important part of the food chains of just about every ecosystem!

Worms can eat half to all of their own weight in food every day, and that creates a lot of castings.

Although they are unable to see or hear, earthworms are sensitive to light and vibrations.

More than seven million people live in Hong Kong; high-rise apartments help provide everyone a place to live.

The people, the worms, and the vegetables all work together to create an ecosystem right there on a little patio 20 stories up in the air. Even in this big city of Hong Kong there are ecosystems, and through them you can see the hand of God at work.

Here is one more thing to consider: Your own body is an ecosystem!

Your body is home to trillions of microorganisms. There are even more of these living on the outside of your body than the number of people living in the world right now! And even more than that live inside of you. The microorganisms that live inside your stomach help you digest the food you eat. There are also microorganisms that live in your nose and help protect you against infections. All these microorganisms inside and outside of you work together in different ways and form a type of community. These communities are all part of the human ecosystem, and together they make up what is called the human microbiome. You have a very specific microbiome; every person has his or her own unique ecosystem!

There are large and small ecosystems all around you. They are in your neighborhood, on your farm, or even in your city.

By taking the time to look around and enjoy these ecosystems, you can better appreciate this amazing world you live in. It will also help you feel grateful for how much God has done for you. He truly knows about the sparrows, the lilies, and you!